QUICK GUIDE TO CULTIVATING ORGANIC MUSHROOMS

Mastering the secrets of organic mushroom cultivation

WALTER K. BYRNE

INTRODUCTION 1
CHAPTER 1 1
Life Study of Organic Mushroom 1
CHAPTER 2 1
Best Mushrooms to Grow 1
CHAPTER 3 1
How to Grow the 5 Easiest Mushrooms 1
CHAPTER 4 1
Comprehensive Understanding of the Growing Process of Organic Mushroom 1
CHAPTER 5 1
Essential Equipment for growing mushroom 1
CHAPTER 6 1
How to Prevent Some Problems that May Occur 1
CHAPTER 7 1
Preparing and Processing Guide 1
TABLE OF CONTENT 1

All rights reserved. No part of this publication may be reproduced, distributed, or transmitted in any form or by any means, including photocopying, recording, or other electronic or mechanical methods, without the prior written permission of the publisher, except in the case of brief quotations embodied in critical reviews and certain other noncommercial uses permitted by copyright law.

Copyright © Walter K. Byrne, 2023.

INTRODUCTION

Organic mushrooms have become increasingly popular in recent years, as more and more people seek out healthier alternatives to traditional foods. But what are organic mushrooms, exactly? In this book, we'll explore the world of organic mushrooms and discuss why they might be a healthy addition to your diet.
So what makes an organic mushroom "organic"? It means that the mushroom has been cultivated without the use of pesticides or other chemicals., meaning these mushrooms contain only natural ingredients. Furthermore, there is no risk of contamination with toxins such as heavy metals or radionuclides like certain synthetic fertilizers can produce when used on conventional crops; whenever you eat an organically grown crop (like any kind of vegetable) you reduce your exposure to potential health risks caused by pesticide residue!
Organic mushrooms also provide nutritional benefits over their conventional counterparts: due to their low level of nitrogen-based compounds compared with conventionally farmed fungi — known as nitrates— they offer increased

levels of vitamins and minerals while containing much less saturated fat than regular button varieties do. And because all parts of the mushroom contribute solely to its nutrient content, each variation boasts numerous important dietary components!

Organic Mushrooms may be worth looking into for anyone curious about experiencing nature.

Are you looking to start your own mushroom-farming business? Organic mushrooms can be cultivated in just about any environment, indoors or outdoors. They require minimal investment and offer a great return on investment. There's also the added benefit of being able to enjoy fresh organic mushrooms right at home!

In this book, we will discuss everything that you need to know before getting started with organic mushroom cultivation including setting up an indoor growing area, its benefits over conventional methods, and our tips for a successful harvest. We'll also answer some common questions that many aspiring cultivators have when starting. So let's get into it!

CHAPTER 1

Life Study of Organic Mushroom

Organic mushrooms are a type of mushroom that is grown without the use of any pesticides or artificial fertilizers.

They can be found in grocery stores, farmer's markets, and specialty stores.

Organic mushrooms are a good source of dietary fiber, protein, and vitamins. They also contain antioxidants, which can help protect the body against disease. Organic mushrooms can be eaten fresh or cooked. They can be added to salads, soups, and other dishes.

Organic mushrooms are available in a variety of colors and sizes. They can be white, brown, yellow, or black. They can also be small or large.

Organic mushrooms are a healthy and delicious addition to any diet.

Since the beginning of recorded history, people have eaten mushrooms; the Greeks thought they gave warriors vigor in combat, and the Romans thought of them as "Food of the Gods." Mushrooms have long been revered in Chinese culture as a wholesome dish and an "elixir of life." They have been a part of human society for countless years, and because of their sensory qualities, they have attracted a great deal of interest in the most significant civilizations in history.

They are also known for having appetizing culinary qualities. Because they are minimal in calories, carbs, fat, and sodium, as well as being cholesterol-free, mushrooms are a valuable food in today's society. In addition, mushrooms offer significant amounts of protein, fiber, selenium, potassium, riboflavin, and vitamin D. Mushrooms have a long history of use as food, but they are also valued for their healing abilities and qualities in traditional medicine. It has been claimed to have positive health effects and be effective in treating some ailments. Mushrooms are said to have a wide range of nutritional benefits, including the ability to treat or prevent conditions including Parkinson's, Alzheimer's, hypertension, and a higher risk of stroke. Due to their antitumoral properties, they are also used to lessen the likelihood of cancer invasion and metastasis. Mushrooms are essential sources of bioactive chemicals and also have antimicrobial, immunological system, and cholesterol-lowering properties. Some mushroom extracts are utilized to support human health and are available as dietary supplements as a result of these qualities.

Worldwide, gourmet cuisine has used mushrooms as an ingredient because of their distinctive flavor and the value they have historically held in the eyes of mankind as a gastronomic marvel. Although there are more than 2,000 species of mushrooms in the world, only around 25 are commonly consumed as food and only a few are grown for commercial purposes. Mushrooms are regarded as a delicacy with excellent nutritional and functional value and are also recognized as nutraceutical foods. Because of their organoleptic quality, therapeutic potential, and economic relevance, mushrooms are of great interest.

Nevertheless, it can be difficult to distinguish between edible and medicinal mushrooms because many commonly consumed culinary species also have medicinal benefits.

Agaricus bisporus is the fungus that is most often farmed, followed by Lentinus edodes, Pleurotus species, and Flammulina velutipes. The output of mushrooms continues to rise, with China being the world's largest producer. However, due to their beneficial nutritive, sensory, and particularly pharmaceutical properties, wild mushrooms are becoming more significant.

In addition to some primary metabolites like oxalic acid, peptides, and proteins, mushrooms could potentially be a source of new antimicrobial chemicals, primarily secondary metabolites including terpenes, steroids, anthraquinones, benzoic acid derivatives, and quinolones. The most researched species, Lentinus edodes, appears to exert antibacterial effects on both gram-positive and gram-negative bacteria.

They offer exceptional nutritional value since they are low in fat but high in essential fatty acids, relatively rich in protein, and contain considerable amounts of fiber and necessary amino acids. Additionally, edible mushrooms have a high nutritional value in terms of vitamins (B1, B2, B12, C, D, and E). Because of the synergistic effects of all the bioactive components present, they may be a wonderful source of many different nutraceuticals and may be utilized directly in human diets to boost health.

Due to their immunomodulatory and anti-cancer characteristics, a wide range of mushrooms has been used traditionally in many different cultures for the preservation of health as well as the prevention and treatment of diseases. The interest in mushrooms' potential as pharmaceuticals has risen significantly in the past 10 years, and it has been hypothesized that many of them function as miniature pharmaceutical factories that create substances with amazing biological capabilities. Additionally, novel medications against aberrant molecular and biochemical signals that cause cancer can now be developed thanks to the increased understanding of the molecular basis of carcinogenesis and metastasis.

Mushrooms and fungi have more than 100 medicinal properties, with antioxidant, anticancer, antidiabetic, antiallergic, immunomodulating, cardiovascular protector, anticholesterolemic, antiviral, antibacterial, antiparasitic, antifungal, detoxification, and hepatoprotective effects being the most important ones. They also can prevent tumor development and inflammatory processes. It is known that many macrofungi produce bioactive molecules, including polysaccharides, proteins, fats, minerals, glycosides, alkaloids, volatile oils, terpenoids, tocopherols, phenolics, flavonoids, carotenoids, folates, lectins, enzymes, ascorbic acid, and organic acids in general. These bioactive molecules can be found in fruit bodies, cultured mycelium, and cultured broth. In terms of modern medicine, polysaccharides are crucial, and -glucan is the most well-known and adaptable metabolite with a wide range of biological activity.

A healthy, balanced diet is the best defense against disease, particularly oxidative stress. Mushrooms have a long history of usage in oriental medicine for both disease prevention and treatment. These days, mushroom extracts are sold as dietary supplements because of their benefits, particularly for boosting the immune system and anticancer activity. In this study, we looked at the nutritional value, chemical makeup, nutraceutical content, and marketability of the most widely grown edible mushrooms on earth.

Edible mushrooms have a high protein, fiber, vitamin, and mineral content as well as low-fat levels, which contribute to their nutritional value. Due to the fact that they contain all the required amino acids for adult requirements and have a higher protein content than most vegetables, mushrooms are highly beneficial for vegetarian diets. Additionally, edible mushrooms have a wide range of bioactive substances with a variety of advantages for human health.

It is significant to note that the chemical makeup and nutritional content of edible mushrooms may be influenced by the growth features, stage, and postharvest state. Significant variances can be found both within and across species. The approximate moisture content of mushrooms is high and ranges from 80 to 95 g/100 g. As previously indicated, edible mushrooms contain 200–250 g of protein per kilogram of dry matter; the most abundant amino acids are leucine, valine, glutamine, glutamic, and aspartic. Due to its low-fat content—20–30 g/kg of dry matter, with the primary fatty acids being linoleic (C18:2), oleic (C18:1), and palmitic (C16:0)—mushrooms are low in calories. Ash content in edible mushrooms ranges from 80 to 120 g/kg of dry mass (mainly potassium, phosphorus, magnesium, calcium, copper, iron, and zinc). Eating mushrooms has a lot of carbohydrates, including chitin, glycogen, trehalose, and mannitol.

They also have fiber, -glucans, hemicelluloses, and pectic materials. In addition, farmed edible mushrooms contain large levels of the sugars glucose, mannitol, and trehalose, but little fructose or sucrose. With high levels of riboflavin (vitamin B2), niacin, folates, and traces of vitamin C, B1, B12, D, and E, mushrooms are also a good source of vitamins. Since mushrooms are the only naturally occurring non-animal food source of vitamin D, they are the only vitamin D source suitable for vegetarians. Contrary to cultivated mushrooms, which are typically produced in darkness and require UV-B light to create vitamin D2, wild mushrooms are typically great sources of vitamin D2.

Edible mushrooms include nutrients, but some have also been proven to contain significant levels of bioactive substances. Edible mushrooms can contain a wide range of biologically active compounds, and the amounts of these compounds are influenced by variances in the strain, substrate, cultivation, developmental stage, age, storage conditions, processing, and cooking techniques.

The secondary metabolites (acids, terpenoids, polyphenols, sesquiterpenes, alkaloids, lactones, sterols, metal chelating agents, nucleotide analogs, and vitamins), glycoproteins, and polysaccharides—primarily -glucans—are the different types of bioactive compounds found in mushrooms. Additionally, new proteins with biological activities have been discovered, including lectins, proteases and protease inhibitors, ribosome-inactivating proteins, hydrophobins, and lignocellulose-degrading enzymes. These proteins can be used in biotechnological processes and for the creation of new drugs.

Numerous edible wild mushroom species, including Tricholoma matsutake, Lactarius matsutake, and Boletus aureus, are valued in China both as food and in traditional Chinese medicine. Many wild-grown mushrooms are regarded as healthy food for the consumer and are almost comparable to meat, eggs, and milk due to their abundance of proteins, carbs, necessary minerals, and low energy levels.

A variety of bioactive polysaccharides or polysaccharide-protein complexes derived from medicinal mushrooms seem to improve innate and cell-mediated immune responses as well as demonstrate anticancer effects in both animals and people. Numerous of these mushroom polymers have been found to have immunotherapeutic effects in the past by promoting tumor cell growth suppression and death. In Asia, a number of mushroom polysaccharide compounds are widely and successfully utilized to treat various cancers and other disorders after advancing through clinical studies. Selected mushrooms are thought to have 126 different therapeutic effects.

The most well-known and effective mushroom-derived compounds with anticancer and immunomodulating activities are polysaccharides. The following specific carbohydrates with these properties have been quantified in various mushrooms: rhamnose, xylose, fucose, arabinose, fructose, glucose, mannose, mannitol, sucrose, maltose, and trehalose. Information on mushroom polysaccharides has been gathered from hundreds of different species of higher basidiomycetes.

The anticancer polysaccharides extracted from mushrooms are acidic or neutral, have potent antitumor effects, and have chemical structures that vary greatly. Glycans with antitumoral properties range widely, from homopolymers to extremely complex heteropolymers. In other words, mushroom polysaccharides do not directly kill tumor cells; rather, they have an anticancer effect by triggering the immunological response of the host organism. These substances reduce bodily stress and may result in a 50% reduction in tumor size while also extending the lifespan of tumor-bearing animals.

CHAPTER 2

Best Mushrooms to Grow

You've decided to produce mushrooms this season, but now you need to figure out how. Perhaps you've already decided where in your farm or garden to put your logs, but what about deciding the kind of tree to use?

Because they are adaptable animals, mushrooms can develop on a wide range of trees. On deciduous hardwood trees, the majority of edible mushrooms as well as all but one of North Spore's strains grow. Coniferous wood is not advised unless you're growing Hemlock Reishi (Ganoderma tsugae). Some members of the Pine family of trees can support the growth of the Italian oyster (Pleurotus pulmonarius), but flushes will be smaller or less frequent than on hardwoods.

Most kinds of deciduous trees can support the growth of mushrooms, however, some are better suited than others for mushroom farming. Each form of mushroom has a preferred kind of wood, therefore pairing the right mushroom with the right kind of log will result in a higher or more reliable yield. Because they are very dense and provide sufficient nourishment for a prolonged, sustained fruiting phase, oaks, and hard maples are recommended wood types for the majority of mushroom species.

Poplars and other soft hardwoods will colonize more quickly and start producing mushrooms sooner, but they typically yield less and don't produce for as long. Oyster mushrooms, however, will thrive on poplars and aspens rather than oaks or maples.

We suggest you use whatever wood is most readily available to you, however, the list below serves as our general recommendation based on North Spore's specific strains. Also, don't be scared to attempt a variety of species or ones that aren't on the list. There are still lots more combinations to try and study! You might see a range of yields, but you might be surprised by the fungal kingdom's adaptability.

Make sure you have access to fresh wood after choosing the type of tree to immunize. Within four weeks of cutting, logs must be infected. If you wait any longer, other fungi that have already begun colonizing the log will have to outcompete your mycelium.

Before You Start: Setting up Your Growing Area

Organic mushroom cultivation requires temperate conditions for optimal growth so if you plan on cultivating them outdoors then make sure the area is protected from weather extremes (e.g., direct sunlight). If opting for an indoor growing setup, here are a few things that you should keep in mind:

• Temperature: Mushrooms prefer temperatures between $18^{\circ}C$-$24^{\circ}C$ during their fruiting cycle and $21^{\circ}C$-23° C during their vegetative phase, which usually lasts 2 months – but could take up to 6 - 8 weeks depending on species type.

• Humidity Level: To maintain adequate humidity levels while cultivating mushrooms inside opt for relative humidities around 70% - 80%. During the fruiting stage, slower mushrooming means lower humidity needs; however, as soon as possible moving towards 90%-95% RH higher moisture helps increase both yield quality & quantity particularly carbon dioxide levels impact yields significantly more than oxygen.

• Substrate Types & Forms: Depending on what type of substrate will determine how stable they remain to colonize by mycelial thread networks allows an easier understanding of specific requirements that must be met timing become familiar with proper sterilization techniques remember there substantial differences among types of grain/rice flour/coir, etc.

Benefits of Organic Mushroom Cultivation vs Conventional Farming;

There are numerous advantages associated with going down the route of organic agriculture versus less sustainable approaches such as traditional monoculture farming or mass chemical usage strips lands microorganisms balance exchanges within one organism can respond to other organisms creating ecosystems and beneficial relationships highlighting healthy communities best practices example used to promote small scale integrated pest management systems.

The Top Mushrooms for Indoor Growing

Do you have any questions on how to begin cultivating mushrooms? There are a few factors to consider before making a choice, regardless of how experienced you are with mushroom cultivation or how new you are to it.

If you're just starting, grow oyster mushrooms first. Mycelium from oyster mushrooms grows quickly and can withstand a variety of temperatures. Most people agree that one of the simplest types of homegrown mushrooms is oysters.

Shiitake mushrooms growing on logs, oyster mushrooms, and shiitake mushrooms

Shiitake is an additional preferred option. Shiitake mushrooms are delicious and simple to grow. If you're just starting with log cultivation, they're fantastic.

The time of year you plan to begin and conclude your endeavor is something to consider for other mushrooms. Matching your mushroom to its favorite season will offer you the finest results because different mushrooms fruit in different seasons.

Other things to think about are your strategies and resources. Do you plan to grow mushrooms on compost, paper, or straw? It is advantageous to be aware of which substrates particular species like before you start.

We'll go into greater detail about selecting the appropriate season and considering your materials below. We'll wrap up with some information on mycorrhizal mushroom cultivation.

Does that seem like a lot to consider? Do not feel overwhelmed. The most important thing is to have fun, and with a little planning, you can grow excellent mushrooms. The first step is learning which kinds of mushrooms to grow at home.

Your decision over which species to grow might be greatly influenced by the season. Knowing that different varieties of mushrooms bear fruit at various periods of the year will help you plan your schedule.

If you're growing mushrooms outside, consider the type of project you'll undertake and how long it will take:

A summer or fall mushroom would be an excellent option for an outside straw bed that was planted in the early summer and was ripe by the end of the season.

You should be aware that morel mushrooms bear fruit in the spring if you intend to grow them. Therefore, it's crucial to know where the ground freezes and to start your patch the previous year.

So you can see how you need to make plans for the fact that the season determines when your mushrooms will fruit.

When growing mushrooms indoors, your reliance on Mother Nature is reduced, but you still need to create the ideal environment. For instance:

Reishi mushrooms bear fruit in the summer, indicating that they prefer a warm environment. The temperature of the space they're in should be adjusted properly.

•Spring is when oyster mushrooms fruit, as you could have imagined. They are a fantastic alternative for chilly locations because your growing space doesn't need to be as warm.

The best times to cultivate certain common varieties of mushrooms indoors are listed below. Although not exhaustive, this list is a decent place to start.

The term "season" is relative depending on where you reside, as a last point. Your summer could be brief. Or, you can have winters when it never drops below 40 degrees (lucky you!). Use this list as a basic reference while keeping your neighborhood in mind.

Spring
•Shiitake mushrooms (from spring through fall)
•Turkey tails and morel mushrooms (from spring through winter)

Mid-Summer

- Reishi mushrooms (and many other polypores) •Elm oysters •Garden Giant (Wine Caps) •Shiitake mushrooms •Chicken of the woods

Fall and the end of summer
- Maitake mushrooms
- Pioppino
- Lion's mane
- Button mushrooms (different Agaricus species)

Winter and late fall
- Cordyceps
- Enokitake
- Shaggy mane

Which Technique is Best for Growing Different Mushroom Types?

Another thing to think about is how you grow the mushrooms. While some mushrooms grow more quickly and easily with straw, some do better on wood. The greatest mushrooms to grow at home truly depend on the materials, time, and space you have available.

Despite this, there aren't any absolute laws dictating how to grow specific varieties of mushrooms. However, when choosing a species, keep in mind your approach and the materials you'll be using. Your ability to grow mushrooms will depend on how much time, money, and effort you have to invest.

Logs/stumps :

Some species make excellent candidates for growing on stumps and logs. Despite taking 6 to 12 months to colonize, fruiting can last for years. Oh, and they taste great, too:

Turkey tails, lion's mane, shiitake, maitake, reishi (and other polypores), several oyster species, and wood ears.

Shiitake is an excellent option if you're just starting with log cultivation. They are simple, adaptable, and not overly particular about the type of wood they prefer. Because maitake and reishi are more challenging, you might want to save them for when you are more skilled.

Chipped wood beds:
While some varieties of mushrooms prefer wood, logs are more difficult to produce than wood chip beds:
- Giant plants and wine caps
- unkempt mane
- On wood chips, an unkempt mane is developing.
- Outdoor beds made of straw:

Several species will flourish on an outside bed of soil, organic matter, straw, etc.

Various varieties of oyster mushrooms
- Agaricus genus
- Garden huge morels and wine caps
- unkempt mane
- Enokitake\sPioppino

Remember that these are recommendations only. Straw works just as well as wood for growing shiitake mushrooms. On wood chip beds, you can also experiment with enokitake.

This list merely illustrates how your preference for mushrooms may influence your material selection and vice versa. It pays to do some research on the optimum growing practices for the mushroom you have selected.

If time and resources are limited, consider using mushroom growing kits to make delightful delicacies. Read this article to learn more about mushroom kits.

All you need to do now is choose your species because the majority of the labor-intensive task of preparing the spawn and substrate has already been done for you. They have a wide variety, including shiitake mushrooms, oysters, reishi, and other foods.

The Truth About Mycorrhizal Mushroom Production

When it comes to the discussion of mycorrhizal fungi, we have saved some of the best for last. Unfortunately, some of the most delectable varieties of mushrooms are also the hardest to grow—or, at this point, are completely unattainable. I'm talking about a few of your favorites:
- Porcini.
- Matsutake Black Trumpet Chanterelles.
- Black trumpet Matsutake King Bolete (Porcini) Chanterelle.

What distinguishes these mycorrhizal mushrooms? Their mycelia interact in a favorable symbiotic way with plant and tree roots. As a result, the plants receive more moisture and nutrients, and the mycelia have access to carbohydrates.

The hobby grower finds it difficult to replicate this intricate symbiotic interaction since mycelium and a host cannot be forced to communicate in a particular way.

Inoculation Of Test Trees With Mycorrhizal Species.

Although impractical to attempt indoors, there are several techniques you can try outside to inoculate a host tree:

Spores should be injected right into the roots.

Make a spore slurry and dispense it around tree bases. You may find directions for doing this here.

Young tree seedlings should be planted close to an area where mycorrhizal fungi are already naturally fruiting.

You'll need the correct species of tree, some perseverance, and a lot of luck for these approaches to succeed. Some mushrooms are challenging to grow, but not impossible. The effort to do some fast trials is worthwhile!

It's hoped that this chapter has given you more information to consider when picking the kinds of mushrooms to grow. Decide now, and start developing!

CHAPTER 3

How to Grow the 5 Easiest Mushrooms

It's simple to understand why growing mushrooms at home have grown popular. But is it simple to cultivate mushrooms at home? Yes, it is the answer.

Mushrooms are one of the easiest and most profitable crops for even novice growers because they have a high yield and can fit in small places.

Mushrooms can be produced in a variety of ways at home, with some species being simpler to grow than others.

Additionally, there are some growing techniques that need so little work (or knowledge) that they can be used by anyone. We're going to give you some tips on how to grow the simplest mushrooms today. Let's start now!

Three Arguments for Growing Mushrooms

1. *They Are Simple to Grow*

Although it's simple to become lost in the mycology rabbit hole, even amateurs may successfully grow mushrooms.

There's no need to be afraid of new language or approaches (inoculating substrates, anyone?). Starting with mushrooms is far more enjoyable than you may imagine.

2. *Mushrooms Take Up Minuscule Space*

There is enough room in each household, whether it be a house or an apartment, to cultivate mushrooms.

Depending on the kind, you can grow mushrooms in your own house by setting up a Low Tech Mushroom Farm or by growing them in outside garden beds.

3. *Mushrooms Bring in Money*

Mushrooms take relatively little maintenance and grow swiftly and densely.

Assume you have around 10 meters (32 feet) to devote to your mushroom setup, and you have about 10 hours to devote to farming. Every week, you might produce at least 22lbs (10kg) of oyster mushrooms.

In the time that most people spend watching TV each week, you might be producing tons of mushrooms!

What Mushrooms Are the Easiest to Grow?

The mushrooms that are simplest to grow are:
- Mushrooms from oysters.
- Shiitake mushrooms.
- Wine cap fungi.
- Portobello Mushrooms.
- Lion's mane fungi.

Observe what makes these mushrooms so simple to grow:
1. *Mushrooms from the oyster*

The easiest type of mushroom for beginners to grow at home is the oyster mushroom. They come in a wide range of kinds, such as king oysters, pink, blue, and golden oysters, which are all stunningly colorful.

(They are also the variety that our GroCycle Mushroom Kit allows you to grow.)

How Come Oyster Mushrooms Are So Easy to Grow?

The simplest mushrooms to grow are oyster mushrooms since they thrive on a variety of substrates, including free coffee grounds that you can generally get from a nearby cafe.

Additionally, they develop remarkably quickly with minimal effort on your part and are very resilient to competing microbes like blue or green mold.

You can have some fun and try growing oyster mushrooms on a book because they are very tough.

What Advantages Do Oyster Mushrooms Offer?

The following are just a few advantages of oyster mushrooms:
- A trustworthy protein source
- Contains selenium, potassium, phosphorus, zinc, iron, and zinc.
- Ability to modulate cholesterol

2. Shiitake *Mushrooms*

Shiitake mushrooms have long been a staple in Asian cuisine and are now more widely available throughout the world because of their delectable flavor and longer shelf life. They are also regarded as therapeutic mushrooms and have been utilized for thousands of years in traditional Chinese medicine.

Why are shiitake mushrooms such a simple plant to grow? Shiitake mushrooms are suitable for a variety of setups and are simple to grow because they can be grown both indoors and outdoors. They can be cultivated on logs kept in the shade if you have space outside, and they will keep bearing fruit for years after their initial flush. The simplest way to produce shiitake mushrooms is with this technique.

However, some Shiitake strains can also be grown on pasteurized straw, which is the simplest approach for beginners growing Shiitake indoors. Shiitake is most frequently produced on sterilized supplemented sawdust when grown indoors.

What Advantages Do Shiitake Mushrooms Offer?

The following are just a few advantages of shiitake mushrooms:
- Significant vitamin D source
- Possibilities for enhancing immunity
- Contains beta-glucans, a natural anti-inflammatory.
- A lot of antioxidants.
3. Mushrooms with a wine cap

Wine caps are ground-based mushrooms, as opposed to tree mushrooms like oysters. They are therefore perfect for growing in outdoor garden beds with good compost. They are frequently utilized in permaculture systems because they quickly decompose organic debris and eliminate soil-based diseases in garden beds. They are also known as King Stropharia or the Garden Giant.

Why are wine cap mushrooms such a simple plant to grow?
Wine cap mushrooms are simple to grow because they are tenacious outdoor enthusiasts.
They grow aggressively and spread quickly, so your growing efforts will pay off greatly.
They can thrive on a variety of substrates, such as wood chips, sawdust, straw, and leaf litter, making them ideal for making a garden bed out of a variety of diverse substrates.
What Advantages Do Wine Cap Mushrooms Offer?
The following are just a few advantages of the wine cap mushroom:
- creates fertile soil for gardens; a good source of fiber

- more vitamin D as a result of growth circumstances outside
- Pleasant, mild flavor
4. Portobello Mushrooms

Pioppino, also known as the black poplar mushroom, is quickly gaining popularity as a mushroom variety among chefs and is a popular fungus to grow.

It grows wild and is frequently foraged in southern Europe, where poplar trees are abundant nearby.

Why are pioppino mushrooms such a simple crop to grow? Because they may be grown on pasteurized straw indoors or in wood chip or straw outdoor mushroom beds, pioppino mushrooms are simple to grow.

The ideal conditions for indoor growing are low temperatures—below 15°C (59°F)—and high humidity. Without a specific regulated growing environment, this can be a little trickier to do, so we advise that the simplest way to produce pioppino is in outdoor beds composed of wood chips or straw.

When raised in this manner, all that is required is to inoculate a bed or patch with Pioppino spawn and wait for it to bear fruit in the fall when the conditions are ideal.

What Advantages Do Pioppino Mushrooms Offer?

The following are just a few advantages of the pioppino mushroom:
- packed with antioxidants
- gives pasta, risotto, soups, and noodle dishes more texture.
- The high linoleic acid content
- a wealth of nutrients; high

5. Lion's Mane Mushrooms

It's obvious how this fungus gained its name because the fruit resembles a shaggy lion's mane.

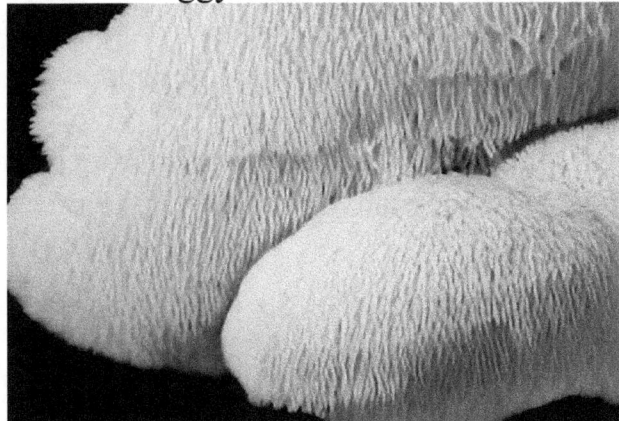

The health benefits of the Lion's Mane mushroom have caused its popularity to soar recently. It is an adaptogen frequently used to improve cognitive function.

What Makes Growing Lion's Mane Mushrooms So Simple?

Lion's mane is among the easiest mushrooms to grow because it develops quickly and bears fruit without being very picky.

Because it thrives at room temperature and grows well from little holes drilled in the growth bag, there is less need for strict temperature control.

What Advantages Do Lion's Mane Mushrooms Offer?

The following are a few advantages of lion's mane mushrooms:
- reduces inflammatory response
- improves gut health and brain function
- used in both cooking and medicinal

How to Easily Grow Mushrooms

We're going to concentrate on the simplest methods for selecting and cultivating mushrooms since, after all, why burden yourself with more work than is necessary?

Where Can I Find a Mushroom Farm?

There are numerous options for your Low Tech Mushroom Farm, and one of them is sure to fit your available area. Yes, even in a condominium!

In this chapter on how to set up a low-tech mushroom farm, I discuss the many alternatives. Here are some suggestions:
- In a cellar
- Your home's extra room
- Inside a container for shipping
- The garage
- In your backyard
- In a wooded setting

Even the container you grow them in can be creative; try using an old laundry basket, a converted bookshelf, or a recycled bucket! Depending on the kind, the mushrooms featured in this book can be cultivated in bags, on logs, in a standard garden bed, or even in jars.

Depending on the method and mushrooms you select, you'll need different things to produce mushrooms.

Generally speaking, you'll need to acquire the following equipment:
A fruiting chamber, or the necessary materials to build one
Shelf for your grow room, if you want to maximize the use of the available area
Substrate (learn about several possibilities in our Complete Guide to Mushroom Substrates) (read about different options in our Complete Guide to Mushroom Substrates)
You can also purchase pre-inoculated substrate instead of mushroom spawn to save time.
Your preferred growth medium, such as buckets, logs, jars, garden beds, or something recycled
Don't forget to take into account the various factors you'll require, like light, temperature, and humidity.
GroCycle Mushroom Kits: Oyster Mushrooms growing from coffee grounds
What is the Most Simple Method of Growing Mushrooms?
Using a ready-to-grow mushroom kit is the simplest way to cultivate mushrooms. These are pre-inoculated and ready to harvest, so they don't need any specific equipment or growing conditions.
Additionally, they may reside on your kitchen bench!
Growing oyster mushrooms on straw or sawdust pellets is another simple way you might attempt if you're feeling a little more secure.
You will require:
- Sawdust pellets, straw, or chopped straw
- spawn of oyster mushrooms
- Develop bags
- A sizable mixing vessel for your substrate
- Scissors
- A water-filled spray bottle

Are you prepared to begin? What you need to do is as follows:

Soak the substrate in cold water in a container, cover it, and set it aside for 30 minutes, or until the substrate has absorbed all the water. The ideal pellet-to-water ratio is 1:1.5.

After the water has been absorbed, top off with your mushroom spawn. For every pound of substrate you use, you should add 10% spawn. Completely combine.

Your spawn and substrate mixture should fill the bags up to two-thirds. The mycelium can develop because of the gap's ability to exchange air. You can fill the bag if you're using chopped straw rather than pellets.

For two to three weeks, seal the bag and keep it in a warm environment of around 20 to 24 F, or 68 to 75 F.

Your bag should appear bright white after two to three weeks since mycelium has thoroughly colonized it and it is now prepared to begin fruiting. To ensure that there is no room at the top of the bag, cut a cross in the side with scissors and tape it down. Spray plenty of water into the region you just sliced open while pulling back the flaps.

Place the bag on a sunny, warm windowsill and mist it twice to three times daily.

In around two weeks, your mushrooms will be ready for harvest. The caps ought to be fully expanded but not flattened just yet. To disconnect from the bag, simply twist.

For an example in real life, see our video on how to cultivate mushrooms at home:

The life cycle of a mushroom

You don't have to strive to manufacture your own spawn; you may simply purchase it. This implies that all you need to do is add inoculated substrate to your bags or bed, cut air holes, water appropriately, and then harvest the mushrooms as they fruit.

Which mushrooms are the most profitable but easiest to grow?

Shiitake and oyster mushrooms are the most profitable and straightforward way to cultivate mushrooms.

Even though you might be tempted to cultivate more precious and difficult types, you must realize that doing so will take more time and money.

The extra time and resources needed to develop them will result in you losing money even though you might make more money at the market.

Oyster mushrooms have the quickest rate of growth.

Oyster mushroom mycelium will completely colonize substrate bags in approximately 10 to 14 days.

Seven days after fruiting, oyster mushrooms will then be ready for harvest.

Following harvest, fresh crops will sprout every 7 to 14 days!

Which mushrooms ought I to cultivate?

Depending on your growing area, time, finances, and needs, you should grow the mushrooms that are most convenient for you or that will serve your needs.

Are you raising mushrooms for their gastronomic and medicinal properties? Pick a selection that appeals to you.

Want to make money? Choose oyster or shiitake mushrooms, which yield a lot in a short amount of time.

We advise beginning with oysters if you're new to mushroom cultivation and moving on to other types once you've gotten the hang of it.

The most challenging mushrooms to grow or the most sophisticated growth techniques aren't always the best.

As you can see, growing tasty, beneficial for your health, and lucrative mushrooms at home is simple.

Some of the healthiest and most lucrative plants are also some of the easiest to grow.

The easiest mushrooms for you to grow will ultimately depend on your resources and skill level. For more information on the best option for you, see our overview on growing mushrooms inside here.

CHAPTER 4

Comprehensive Understanding of the Growing Process of Organic Mushroom

What you must understand
The most healthy and sustainable food source available is mushrooms. It's not difficult to learn how to produce mushrooms. Mushrooms are simple to cultivate at home because they expand rapidly and effectively, turning plant materials into scrumptious eating mushrooms with little effort or effect on the environment. However, one of the crops that home gardeners seem to overlook the most is mushrooms. You may quickly start cultivating your own mushrooms at home with just a little knowledge of the subject.

If you want to grow excellent mushrooms with a high chance of success and no equipment other than a sharp knife, ready-to-fruit mushroom kits are available in our online store. Mushrooms can be grown with these kits in a few days. Mushroom growing kits are fantastic presents for your friends as well! You may purchase mushroom kits here.

the fundamentals before beginning

Understanding the life cycle of mushrooms is crucial if you want to learn how to grow them. As a spore, the mushroom life cycle begins. Because they are so tiny, individual mushroom spores are invisible to the unaided eye. Spores germinate to produce white threadlike organisms known as Mycelium if they are placed in appropriate growing conditions. Mycelium can spread through any type of organic material, including compost, decomposing logs, and even tree roots in the soil. It appears as a network of white mold. When the mycelium is in a good environment, it produces mushrooms, which then produce reproductive spores and release them into the air to continue the cycle.

Only about 2,000 of the more than 10,000 types of mushrooms are known to be edible. Only a few of the edible kinds can be grown. The three types of mushrooms are saprobic, mycorrhizal, and parasitic. Saprobic mushrooms are the easiest to produce since they require dead organic debris to develop. Oyster mushrooms, shiitake mushrooms, button mushrooms, paddy straw, wood ear, enoki, maitake, and many other varieties that you may have seen at the grocery store or farmers market are among the most popular saprobic mushrooms.

Mycorrhizal mushrooms grow in symbiotic partnerships with trees, making it difficult or impossible to grow them. Examples include porcinis, truffles, morels, and chanterelles. Mycorrhizal mushrooms must grow near living trees because that is where they naturally develop, whereas saprophytic mushrooms can grow on a growing medium such as sterilized sawdust or logs. Mycorrhizal mushrooms cannot be grown in a home garden.

Although some saprobic mushrooms are thought to be slightly parasitic, parasitic mushrooms are typically not grown outside of their natural habitat. For instance, reishi mushrooms are frequently cultivated despite occasionally being seen as parasitic. Read more about "cordyceps" if you want to find out how to grow your own tasty parasitic mushrooms. Now that you have a basic understanding of the various species of mushrooms and their life cycle, let's start growing some of them!

With which varieties should you begin?

Oysters, shiitake, wine cap, and lion's mane are by far the simplest to grow among the widely grown kinds. You can grow these mushrooms on logs, hardwood sawdust, and mulch. Beginners frequently use "ready-to-fruit" kits from the shop that have detailed instructions.

If you wish to grow other mushroom species outside of their natural habitat, you will need to put in some effort. These kinds can be grown, but it takes a lot of work and investigation. It might be better to hold off on trying other species until you have had experience growing some of the kinds suggested here.

Clam mushrooms

On sawdust and numerous other substrates like straw, coffee grounds, and cottonseed hulls, as well as many different species of oyster mushrooms, they will thrive. Because they have a robust, gratifying flavor and texture and are simple to grow, oyster mushrooms are the most often used type of cultivated mushroom. Many individuals grow oyster mushrooms at home with amazing success.

Mushrooms shiitake

Compared to oyster mushrooms, shiitake has a firmer texture and a more flavorful flavor. Shiitake mushrooms can be grown indoors or outdoors on logs or sawdust. Be tolerant! You may expect your logs to start producing mushrooms in around six months.

It's crucial to keep in mind that logs must always remain damp for mushrooms to grow; the wood must never totally dry out.

White Wine Cap Mushrooms

Wine caps are a species that is enjoyable to grow and are frequently grown in compost or mulch piles. Wine caps are frequently compared to meatier portobello mushrooms in terms of flavor.

A valuable mushroom is a lion's mane. They are frequently compared to lobster because of their meaty texture and delicate flavor. On sawdust made of hardwood, lion's mane mushrooms thrive.

The fundamental procedures needed to cultivate mushrooms.

The process of cultivating mushrooms is easy. All you need for mushrooms to grow is substrate material (or mushroom food) and spawn (aka mushroom mycelium), along with some time and space. They don't need potting soil, fertilizer, or even sunlight. Of course, the process will be simpler if you are aware of a few more things

Cultures of Mushrooms (living mycelium)

The kind of mushroom cultures (or spores) that are available to you or the person who is making your mushroom spawn will determine how many different kinds of mushrooms you can grow. You can buy spores and cultures online if you don't already have some. Once you've decided the type of mushroom you want to raise, place your order from one of the many reliable mushroom vendors available online. Mycelium in a petri dish or syringe is the most common form that mushroom cultures take. In our online shop, you may find premium mushroom cultures.

Mussel Spawn

A beginning culture of mycelium from which you will grow mushrooms is known as "mushroom spawn." Grain spawn, sawdust spawn, and plug spawn are the three most typical types of mushroom spawn. If it is utilized to spread mycelium to a new food source, anything with mycelium on it is referred to as a spawn. You can create your offspring using modest amounts of existing cultures if you have the right skills. If not, spawn can be bought online.

Prepare the Basis

The food or growing media used to cultivate mushrooms is known as mushroom substrate. The most crucial aspect of substrate preparation is keeping things clean. Contamination is a simple way to eradicate fungi. The ideal growing medium for the majority of mushroom types is freshly cut, disease-free hardwood logs or sterilized hardwood sawdust combined with various supplements (cottonseed hull, soybean hull, wheat bran, etc.). For some kinds of fungi, substrates made of dung and coco coir are also frequently used. In our online shop, you can buy prepared mushroom substrates.

Inoculation (spawning) (spawning)

Simply placing your spawn into the substrate is known as inoculation; this is akin to sowing seeds. While sawdust substrate is frequently generated by introducing the mycelium in the form of grain spawn, log inoculation entails drilling holes into hardwood logs and filling them with plug spawn or sawdust spawn. Use brand-new spawn for the greatest chance of success.

Incubation (spawn run) (spawn run)

After mixing your spawn into your substrate, wait for it to colonize the entire thing before you start to create the circumstances for fruiting. This is the stage of the growing cycle between inoculation and fruiting. Mycelium will be seen advancing through the substrate. Although it doesn't need sunshine or soil to flourish, mushroom mycelium does best in specific environments. The mycelium may suffer if there is too much carbon dioxide present. The mycelium will dry out and die if it is too dry. The length of the incubation period varies depending on the type of mushroom and the substrate.

Fruiting (cropping) (cropping)

The most exciting element of mushroom cultivation is this. Now that colonization is finished, fruiting can start. Fresh air and colder temperatures typically start this process. The mushrooms start as little 'buttons' or 'pins,' from which they will grow into enormous, stout mushrooms!

Harvesting

Those tiny pins will develop into mature fruiting bodies in a week or two that are prepared for consumption! You now possess your very own mushrooms. You can now call yourself a mushroom farmer.

Where can I find a mushroom farm?

The best location to grow mushrooms is up to you; they may be grown in practically any place. The majority of mushrooms may grow best in environments with high humidity and temperatures between 55 and 75 degrees Fare height. This can be accomplished indoors by getting a small humidifier and placing it close to your growing area. Oyster species, for example, can fruit without additional humidity, but others, like shiitake and lion's mane, thrive when humidity levels are raised.

Indoor mushroom cultivation

The most common mushroom species can be cultivated indoors on a growing medium like sawdust, although oyster mushrooms are the easiest to grow. The sawdust is typically put in a bag and occasionally marketed "ready to fruit" as a kit for growing mushrooms on surfaces like a counter or windowsill. Because mushrooms release carbon dioxide as they grow, ventilation is necessary if you're cultivating them inside. Growing a lot of mushrooms in your home is also a bad idea because the spores can irritate some individuals.

Mushroom cultivation in a garden

On wooden planks outside, a wide variety of mushrooms flourish. Seek out a shaded area that is shielded from wind and rain. The logs must be shielded from the sun's rays and the wind's drying effects while growing mushrooms outside on logs. Additionally, the logs should be kept damp but not drenched. The lifespan of a log is up to three years or longer.

Some mushrooms thrive in mulch as well. For example, straw chips or mulch can be used to grow wine caps. You may cultivate wonderful wine caps in your vegetable garden.

Some people's curiosity will not be satiated by hobby-level mushroom gardening. Commercial mushroom farming may be the solution for these folks, but what does the future hold for these would-be farmers?

Farming while growing mushrooms

Small farms are frequently found in residential or urban areas. Although it might appear challenging, urban mushroom farming has some benefits over other types of industrial-scale mushroom production. The fact that urban farms are situated close to the majority of their target market is by far the biggest benefit. Because of this, they can convey their goods quickly and efficiently from the farm to the consumer's table without having to worry about high shipping prices or lengthy storage periods.

Since the early 1900s, people have been growing mushrooms indoors, starting with what we now know as white button mushrooms. Although there are many different varieties of cultivated fungi available on the market, white button mushrooms are by far the most common edible mushroom in North America. Chefs are becoming more enamored with exotic, gourmet mushrooms like shiitake, oyster, and lion's mane. There is a lot of room for creativity when it comes to commercial indoor mushroom growing. Technology is a driving force in this sector, and entrepreneurs have a promising future.

Although it has been practiced outside for generations, modern advancements have made it more widely available than ever. Shiitake mushrooms and other hardwood tree log-grown mushrooms need a lot of space to grow successfully. Entrepreneurs looking to create a mushroom farm don't have to travel very far because there are many untapped hardwood kinds of wood in North America. Shiitake mushroom cultivation has several advantages, including inexpensive initial costs and a potential for significant return on investment. A lot of businesses are creating new technologies to make it easier for business owners to grow mushrooms outside on logs, which will result in more chances in the future.

The aforementioned images were taken in a condominium complex's backyard, where a modest building is producing up to 50lbs every week during the Alabama winter.

Final Reflections

Growing mushrooms at home is a delightful hobby that can yield a steady supply of fresh food. Mushrooms are intriguing organisms. Ask questions when you first start, don't be shy about it! Online forums are a great place for mushroom growers to assist newcomers. Many mushroom farmers enjoy showing newcomers the ropes.

Although the fundamentals of mushroom cultivation are straightforward, there is still much space for development and innovation in this field. Growing mushrooms is really fascinating because new growing techniques are constantly being discovered. Keep up with new developments and never stop learning.

CHAPTER 5

Essential Equipment for growing mushroom

the supplies needed to grow mushrooms

They are a favorite among nutritionists because of their high protein level, low fat, and low cholesterol content. Nearly all of the amino acids needed by the body are found in mushrooms, and most of their by-products are employed in cosmetics and medications.

The bare minimum required to start growing mushrooms is just five items:

- mushroom spawning
- Substrate (the growth media) (the growing medium)
- Grow Totes (or buckets)
- Thermometer
- Water Sprayer

You might require more materials or fewer resources, depending on how sophisticated your growth operation is.

Depending on how many mushrooms you plan to grow, you'll need different things. moreover, the kind of mushroom you'll be growing and your strategy.

1. **Spawn**

Spawn is one of the most important ingredients you'll need to start growing mushrooms. If you want to grow high-yielding, top-notch mushrooms, make sure the source of your mushroom spawn is reliable.

2. **Substrate**

The mycelium will receive all the nutrients it needs from your substrate to grow and start producing mushrooms. As substrates for mushrooms, materials such as sawdust, cardboard, coffee grounds, coco coir, and others may be utilized.

3. Grow Totes

Your spawn and substrate will need a place to stay while growing mushrooms. Large plastic buckets or grow bags would do. They provide the high CO_2 and humidity levels necessary for mycelium to thoroughly colonize the substrate.

For beginners, growing in bags is an excellent place to start. The biggest benefit of bags is that they are transparent, so you can see what's happening with your substrate. It facilitates the early detection of contamination and other problems.

4. Thermometer

A thermometer is necessary to check the temperature of your mushrooms. Different temperature ranges are favorable for various types of mushrooms. Different varieties of oyster mushrooms prefer various temperatures.

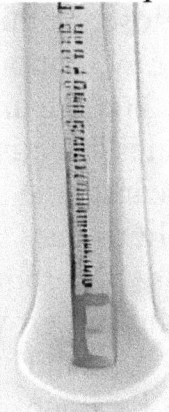

For instance, blue oyster mushrooms like fruiting temperatures between 12 and 18 C. (45-65 F). Temperatures between 18 and 30 degrees Celsius are ideal for pink oyster mushrooms to ripen (64-86F).

5. **Water Sprayer**

You'll require a water sprayer when your mushrooms are ready to fruit. They will remain moist and the appropriate humidity level will be maintained.

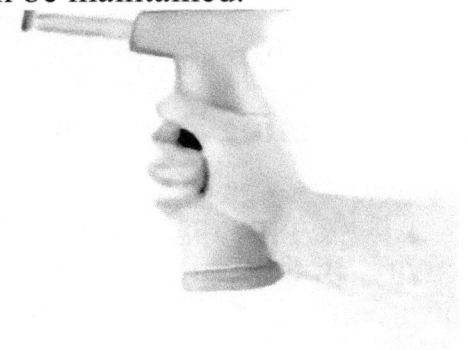

A simple spray bottle with water inside would do if you were only growing a few mushrooms. To prevent them from drying out, you should spray your grow bags at least twice a day.

Materials containing cellulose, hemicellulose, and lignin (such as rice and wheat straw, cotton seed hulls, sawdust [SD], waste paper, leaves, and sugarcane residue) can be used as mushroom substrates because the main nutrients are less nitrogen and more carbon.

One square meter of mycelium may support the growth of 30 kilograms of mushrooms. In summary, 560 m2 of space can support 17 tons of mushroom growth.

Growing mushrooms is a fairly flexible pastime. Mushroom cultivation can be done easily and cheaply at home. To maximize output and productivity, you will need more supplies as your mushroom farm expands in size and complexity.

CHAPTER 6

How to Prevent Some Problems that May Occur

Around the world, growing mushrooms is a relatively common activity. Both small-scale and large-scale farmers cultivate mushrooms. It is also a practical way to work for yourself. Mushrooms have a very rapid rate of growth while also being quite sensitive to their surroundings. As a result, cultivating them might occasionally be challenging. They also have certain challenges, and I will outline some typical issues with mushroom cultivation in this article along with solutions.

Temperature, airflow, and humidity variations can all have a direct effect on how well mushrooms grow. These could cause issues with your mushroom culture such as browning and yellowing, mold, and contamination. To ensure you obtain the biggest harvest possible, it's crucial to give your mushrooms the best circumstances.

Common Issues with Mushroom Growth

What are the main issues in growing mushrooms?

1. *Contamination*

One of the most frequent issues when growing mushrooms as a novice is contamination. Since they are a type of fungus, mushrooms do best in warm, humid settings. Other microorganisms are also encouraged to grow by the same causes. This poses a serious risk when mushroom cultivation is still in its infancy. Other contaminating bacterial and fungal species may also take hold of the substrate in the early stages when our preferred mushroom strain is not well-established in the substrate.

In selective cultures like this, any bacterial or even fungal species that is not your preferred strain is regarded as contamination. Contamination must be addressed as soon as feasible because it may occur in the later stages of growth as well. The growth of your mushrooms may be hampered if the contaminations are not handled correctly. The mushrooms may grow thin or, in some extreme circumstances, not bloom at all because they must compete with the contaminated species for the scarce nutrients in the substrate.

Solution: You should address contamination mitigation at the very beginning of your cultures, and more crucially, at inoculation. It is important to work in a spotless environment and to utilize sterile tools, including your own hands. Any tools you use to produce mushrooms, such as humidity-controlled trays or containers, should also be disinfected.

Cleaning everything with 70% ethanol or isopropyl alcohol is the only foolproof method for accomplishing this. Additionally recommended are hygienic measures including donning a surgical mask and gloves. Additionally, you must confirm that the strain of spawn or spawn you utilize to inoculate the substrate is clean and uncontaminated beforehand. Another crucial step you should perform before inoculation is pasteurizing your growing media.

The substrate can be pasteurized using several techniques.
- Pasteurization in a hot water bath.
- Using cold water to pasteurize lime.

How to sanitize the substrate for mushrooms

Immerse the substrate in boiling water for at least one or two hours when pasteurizing food in a water bath. Put your support in a cold water bath that has been treated with hydrated lime for 24 hours to do a cold water lime pasteurization. This lowers the water's pH and gets rid of impurities. The substrate is put under pressure and exposed to temperatures over 250 degrees Fahrenheit during sterilization.

Any pollutants that might be present in the substrate, both living and dormant, are fully eliminated during sterilization. If you want to carry out mushroom cultures on a large scale, it is advantageous to invest in microbiological equipment such as a laminar flow and an autoclave.

These tools have a history of being reliable and efficient in microbiological procedures. Your growth media won't become contaminated with bacteria or fungus in the airflow if you inoculate it when it is in a laminar flow.

It is also useful to administer the inoculation close to an open flame to lessen contamination from airborne bacteria and fungus. The growth media can be pasteurized with success using an autoclave. On mushroom growth bags, contamination may take the form of brown, black, green, orange, or pink stains or patches.

Smaller portions should be isolated, and the mycelium will naturally remove the contamination while continuing to develop normally. However, it is preferable to remove contaminated bags or pots from the growth space and properly dispose of them if the mold has spread more than 20%.

2. *Insufficient Moisture*

One of the most crucial variables you may need to control in mushroom cultures is the humidity level in the growing chamber. Both low and high humidity levels may be harmful to the mushroom's ability to develop. The mycelium may dry out and eventually perish if the air in the growing chamber is too dry or if the substrate doesn't have enough moisture. Mushrooms typically retain a lot of moisture in their fruiting bodies and consume a lot of water as they grow. If the mushrooms are not adequately irrigated, dry, brittle stems and fractured caps may be seen.

Solution : Constantly check that your substrate is getting enough water. You might also sometimes mist your mushrooms with water. But be careful not to spray too much. An efficient way to guarantee that your growth chamber has precisely the right amount of moisture for the healthy growth of mushrooms is to get a humidifier. Additionally, measures should be taken to reduce water loss from the mushrooms. Therefore, it is preferable to confine the mushroom cultures to a small space.

The growth chamber's airflow pathways should be kept higher and lower until mushrooms are present. To avoid overdrying, the airflow shouldn't immediately touch the level of the oyster mushrooms.

3. *Excessive Moisture*

Although they prefer moist conditions, mushrooms can suffer from excessive humidity. While excessive moisture in the substrate can encourage contaminations, excessive air humidity can result in issues like "Fuzzy feet" or browning and yellowing of the caps.

Solution : Make sure your growth bags effectively drain any extra water and don't collect any puddles.

Make sure you don't water too much. Small amounts of watering should be applied many times per day. Avoid immediately watering the fruiting bodies. The substrate alone should be watered.

4. *Insufficient Fresh Air*

Mushroom development may be hampered by inadequate ventilation. The oxygen supply to the mushrooms may be constrained by insufficient exchange of fresh air. Like ourselves, mushrooms consume oxygen in their metabolism and expel carbon dioxide. The stems will thin out and the caps will get smaller than usual if there is insufficient fresh air flow in the chamber. Additionally, abnormalities like fuzzy feet in the mushrooms may result from a lack of oxygen. These can significantly lower your mushroom harvest's yield.

Solution : Make sure your mushrooms' growing chamber is well aired and that there is constant fresh air exchange. The mushroom growth chamber can be equipped with a fanning mechanism to achieve this. The mushrooms can also be manually brushed several times each day.

5. *Insufficient Lighting*

Mushrooms do not require light to make energy; instead, they break down organic matter to do so. They do, however, require a certain amount of diffused light to thrive. Mushrooms can develop long, thin stems due to a lack of light. Your mushrooms will lean in the direction of the light in the chamber, so you can tell if they don't get enough light. Fuzzy feet in the mushrooms might also be brought on by inadequate lighting.

Solution: Ensure that your mushroom growth chamber receives an adequate amount of natural light. Installing a lamp with gentle light is another option if the existing illumination is insufficient.

6. *Unsuitable Climate*

Some mushroom species may be indigenous to particular climate zones and geographical locales. Because they are so sensitive to their surroundings, each mushroom needs the ideal combination of temperatures, humidity, and ambient light to thrive. As a result, certain mushrooms might not grow in various environments and climates.

Fortunately, you may be able to grow a wide range of mushrooms if you reside in a place with a reasonably temperate climate rather than a farmer who lives in a very hot or cold climate. In contrast, certain mushroom types may even thrive in various weather conditions. Particular attention should be paid by outdoor growers to the best environments for growing the different types of mushrooms.

Solution : If you cultivate your mushrooms outdoors, make sure that the species you choose can flourish in the local environment. In this situation, growing your mushrooms inside might be a preferable choice. Because you may adjust the conditions in your growth chamber to suit the needs of your mushroom.

7. *Incorrect Substrate*

You can choose from a variety of substrates to grow your mushrooms. Traditionally, hardwood pellets, straw, coco coir, coffee grounds, animal dung, compost, and even cardboard have been used to cultivate mushrooms. These substrates can occasionally be used in conjunction with one another. However, some species might have particular substrate needs and won't thrive if given anything different.

For instance, while certain species, like oyster mushrooms, may grow well in a variety of materials, others, like truffles, prefer to grow on the roots of living trees, making it exceedingly challenging to artificially produce them.

Solution : The easiest way to prevent using the incorrect substrate is to educate yourself on the mushrooms of your choice. Some mushroom producers also prefer to give extra nutrients to the mushrooms.

8. *Inadequate Harvesting*

You must harvest your mushrooms carefully and without disturbing the mycelium. The mycelium in your growth bags has the potential to produce further blooms if it is kept intact. The timing of harvest is equally important for mushrooms. Being either early or late could cause issues.

Mushrooms that are picked too soon may still be young and underdeveloped. Even with refrigeration, if you harvest your mushrooms too late, they can go rotten in a matter of days. The mushrooms may also develop aerial mycelium on the fruiting body if you wait too long to harvest them. The mushroom stem, caps, and even the body can become covered with what may appear to be white mold. The texture and flavor of your mushrooms might also change depending on when they are harvested.

Solution: When harvesting mycelium, you should take care not to disrupt it. Holding the stem loosely and twisting the base of the mushroom until it separates from the mycelium is the right method for picking mushrooms. For this, you can alternatively use a set of sterilized scissors or a knife. The ideal time for harvesting a particular variety of mushrooms may vary depending on its life cycle. Oyster mushrooms, for instance, should be picked when the caps open and the rim begins to curl. When the cap opens, the mushrooms are ready for picking. Gills, however, ought to be slightly opened.

9. *Immunized But No Fruit*

This problem may be a result of both inadequate ventilation in the growth chamber and a lack of moisture in the substrate. If not, your particular mushroom variety might be a late bloomer. Each type of mushroom has a distinct growth cycle. Some mushrooms bloom quickly because of their brief lifespans, while others may take up to a year.

Solution:

It could be wise to calmly wait a few more days or weeks at first. Your mushroom can be a slow bloomer and require further time. Second, you can make sure that the growing medium is appropriately moistened and hydrated so that the spawn can grow. To keep your offspring from drying out, you must immediately hydrate the medium after inoculating it.

Lack of adequate air could be another factor in late blossoming. Make sure your chamber has enough fresh air flow if you are preserving your cultures indoors. To do this, a fan can also be installed. Your mushroom cultures might benefit from being moved outside as well.

10. *Mycelium Is Growing But No Fruit*

If your substrate is growing mycelium, which appears as white strands.

There could be a few causes if you observe this but do not experience mushroom blooms despite waiting the recommended amount of time. This problem may be related to faulty spawn or inappropriate media

Solution : It is advantageous to conduct thorough research on your particular mushroom variety to determine when it should bloom precisely and to choose the best substrate type for your mushroom. Make sure to acquire your spawn from a reputable merchant to avoid receiving defective spawns. Wait a few more days if you notice tiny pins along with the mycelium growth; your mushroom may still gradually develop into a complete bloom.

11. *The Substance Isn't Producing*

Each species of mushroom may have unique substrate needs, and they may not thrive when grown on other types of substrate.

Solution: Be sure to identify the best growing environment for your particular species of mushrooms. It is best to conduct a preliminary study in this situation. Make sure you comprehend the timing and the life cycle of your mushroom.

12. *Defective Spawn*

It's crucial to find a nice, viable spawn if you want flourishing mushroom cultivation. If the spawn is poor, no matter how well you execute everything else, the culture as a whole will collapse. The spawn can perish on the way if delivery takes longer or if you don't utilize it right away after it arrives because it doesn't have a lengthy shelf life.

If you keep the spawn for a longer period of time, they may also get tainted.

Solution: Make sure to get your spawn from a reputable vendor who offers quicker shipment. When they come, keep them somewhere cold and dry, and use them right away.

13. *Excessive or Insufficient Spawn*

In your mushroom cultivation, using too much or too little spawn can cause problems. You ought to be aware of the maximum amount of spawn your substrate can support. If you use too little spawn, contamination could occur. On the other hand, if you utilize too much spawn, competition for the substrate's scarce nutrients will impede growth. With excessive spawn use, thermogenesis is also a possibility. Mycelium generates heat when it consumes the nutrients in the substrate. Mycelium death could occur if the substrate heats up too much as a result of too many spawns.

Solution: Do your research and determine how much spawn is appropriate for each unit of the substrate. Depending on the species and the substrate, this may be subjective. Use 2 parts spawn to 8 parts substrate for the tried-and-true ratio. Although this can be a little challenging with some species. But with time and expertise, you might discover the ideal balance.

14. *Quickly deteriorate After Harvest*

The mushrooms may spoil quickly if they are harvested too late. The majority of mushrooms can be harvested while their caps are still dry and partially open. Your mushroom caps may spoil very quickly if they are damp when you harvest them.

Solution : If you collect mushrooms later than usual, use them right away. To determine the right time for harvesting your particular mushroom kind, do some research. Mushrooms are delicate and could go bad if they aren't preserved properly. Mushrooms should be kept in the refrigerator in a plastic bag that is just loosely closed; they shouldn't be securely wrapped in plastic film. There should be ample space for the movement of new air.

15. *Lack of Information*

Many issues can arise if you don't know the variety of your mushrooms. Because mushrooms are sensitive to their surroundings, many things can go wrong if you don't give them the right setting. This can ruin your entire mushroom culture and cost you money, time, and effort.

Solution : Conduct further research. It is the ideal method for comprehending your particular mushroom type. You must have a fundamental understanding of your mushroom's life cycle and the ideal conditions needed for its growth. Information about mushroom culture can be found all over the internet because it is a very popular activity. You might find it useful to speak with a seasoned mushroom farmer as well. You can get assistance from someone with experience troubleshooting any issues you run into.

The challenges listed above are some of the most fundamental and typical ones you could encounter when growing mushrooms. You might not experience all of these problems at once. It's possible that you won't have any of these issues if you fully comprehend your mushroom's needs and meet them on schedule.

Mushrooms are sensitive to their surroundings, and if they are not in the best conditions, many things go wrong. However, if you control the temperature, humidity, ventilation, and substrate moisture at precisely the correct time and in precisely the right amount, you may completely prevent many issues. However, due to their inexperience, novice growers may find this challenging. Therefore, if newcomers want to sustain long-lasting healthy cultures, they should conduct adequate research and have a lot of patience.

CHAPTER 7

Preparing and Processing Guide

You're in for a real treat if this is your first time cultivating mushrooms, which it most certainly was for me. These tasty mushrooms grow quite quickly, so I want you to feel completely equipped and supported as you embark on your mushroom-growing adventure.

Let's begin at the beginning. Someone once said that would be a great place to start.

The following is a step-by-step guide for growing your own organic mushrooms:

1. *Purchase a mushroom-growing medium*

After choosing my new Mushroom Growing Kit, I went home and cleaned my room so that I wouldn't feel self-conscious while taking these photos. Then I decided where I would grow mushrooms.

The Location: I settled on a table close to my west-facing window at the Mushroom Farm. Little Shroomie quickly made several pals because this is where I store the majority of my plants.

2. *Prepare the mushroom-growing medium*
I then looked at the side panel, which had a few tips on how to grow the Mushroom Farm. On the inside, there was also a note with instructions and a tiny spray bottle (super duper cute).

Both sets of instructions advise removing the box's front panel before removing the bag, but I'm a rebel, so I disregarded their advice. Additionally, I believe that when the bag is outside, it is simpler to punch out.

The instructions then tell you to cut an "X" into the bag's front. I did this by carefully holding a pair of scissors open while cutting the shape of an "X" into the plastic covering the top of the dirt block. The flaps ARE NOT to be clipped off! These play a crucial role in aiding the soil's ability to retain moisture by producing a humid, greenhouse-like environment.

I next raised the flaps I had just created by cutting the "X" in the ground and scratching the earth with a fork. To conduct this scratching, you could basically use anything, such as a stick, a spoon, or a magic wand. The secret is to scratch firmly but not so hard that it appears you are cutting through the brick. Simply thoroughly scrape the white substance to make it appear less white, a little less dense, and looser.

It's crucial to remember this step because it's essential for promoting mushroom development.

2. Soak *the Growing Medium for Mushrooms*
Soaking the soil is crucial for soaking it and giving it the moisture boost it needs to support your mushrooms.

I prepared a large pail of water after scratching. I don't know why I thought I would need a multi-gallon heavy-duty bucket for this step, but really you just need a dish big enough for the mushroom bag to lie face down in. Water should be put in the bucket or bowl. The bucket needs to soak for 6 to 10 hours, so place it somewhere out of the way. I decided to soak it all night.

Place the mushroom bag inside the water bucket with the "X" facing down once it has been ready. It will probably float on top of the water a little unevenly. Just be certain that the "X" side is immersed. To be safe, I placed an empty heavy-duty Tupperware container on top of the mushroom bag, although this is not required.

If part of the dirt leaks out while it's soaking, do not worry. This is entirely typical.

3. *Soak and develop the mushrooms*

I put the mushroom bag back into the box and put it back in the location I had chosen near the window with indirect light the next morning (or 6–10 hours later).

It was supposed to be placed with the side bearing the "X" facing up, but I forgot to do that and it just stood straight up. I filled the spray bottle with water from the faucet as soon as I placed the Mushroom Farm box in its location. I carefully opened the flaps of the mushroom kit and sprayed the soil with four or five pumps.

About three times every day, when I remembered that I was growing mushrooms in my room, I pretty much always did this.

For the following eight days, I carried out the spraying procedure every day until I saw young mushrooms beginning to sprout. This process, known as "pinning," usually takes 1-2 weeks to complete.

Pinning up close

The instructions said to stop watering once the mushrooms began to develop. I did, however, continue to spray once a day because my room is frequently quite dry. If it appears that your mushrooms are starting to dry out, do this.

After that, I essentially did nothing but sit back and unwind for the following week while my mushrooms grew from tiny sprouts to full-grown mushrooms. From here, the mushrooms grow so quickly!

Harvest Your Mushroom Crop in Step 4!

The mushrooms will surely double in size daily, just as the box claims. The mushrooms stopped growing after five days following pinning and were about two to two and a half inches in size.

At that time, I removed all of the mushrooms from the growth media by carefully picking them out. I removed the dirt particles that were stuck to the mushroom caps with water and spread some of them out to dry.

I cooked one of my favorite dishes with the remaining mushrooms. Actually, it's heaven on earth. It makes me want bacon more. Seriously.

I melted some unsalted butter in a cast iron skillet over medium heat. I then divided each mushroom cap into individual ones and added them to the butter-coated pan. After about 5-7 minutes of sautéing in the skillet, the mushrooms will start to become a dark golden brown on some of the flatter surfaces. I then added a dash of salt and began to eat the mushrooms. Although it may seem pretty basic and straightforward, this dish is incredibly wonderful. The homegrown Pearl Oyster Mushrooms from your mushroom farm can be used in myriad incredible ways, but this is just my tried-and-true favorite for a straightforward breakfast.

CONCLUSION

In conclusion, embarking on the journey of organic mushroom cultivation opens a world of fascinating possibilities for both enthusiasts and aspiring cultivators alike. Throughout this comprehensive guide, we have delved into the intricate art and science of cultivating mushrooms in harmony with nature, fostering not only bountiful harvests but also a deeper connection to the earth.

Our exploration began with an understanding of the fundamental principles that underpin organic mushroom cultivation. From selecting the right mushroom species to creating optimal growing conditions, we've unraveled the key components that contribute to a successful harvest. By embracing sustainable practices, we not only nurture the growth of mushrooms but also contribute to the well-being of our environment.

The cultivation techniques discussed here go beyond mere instructions; they represent a synergy between traditional wisdom and modern innovation. We've demystified the process of creating nutrient-rich substrates, fostering mycelial growth, and inducing fruiting bodies to emerge. Each step has been carefully crafted to empower you with the knowledge needed to cultivate a diverse array of mushrooms, from the delicate oyster to the robust shiitake.

While the journey of organic mushroom cultivation requires dedication and patience, the rewards are abundant. The act of tending to your mushroom patch becomes a meditation, a way to reconnect with the rhythms of the natural world. As you witness the first pinheads forming and the caps expanding, you'll be reminded of the incredible symbiosis that occurs beneath the soil's surface – a reminder of the hidden marvels that exist beyond our everyday perception.

But this guide goes beyond the technical aspects. It encourages a mindset shift towards sustainability and respect for the environment. By choosing organic practices, we not only produce wholesome mushrooms but also contribute positively to the delicate balance of our ecosystems. Through your journey, you become a steward of the land, nurturing life in all its forms.

As you set forth on your own mushroom cultivation endeavors, remember that every trial and every triumph adds to your expertise. Adaptation and innovation are your allies, as you navigate the intricacies of changing seasons and evolving conditions. Just as mycelium spreads its delicate network underground, so too does your understanding of this captivating practice expand with each chapter of experience.

In the grand tapestry of organic mushroom cultivation, you've now woven your unique thread. May this guide continue to serve as your companion, offering insights, troubleshooting advice, and inspiration whenever you seek it. From the moment you harvest your first cluster of pristine mushrooms to the days when you expertly navigate the challenges of larger-scale cultivation, remember that you're contributing to a movement that champions both delicious flavors and a harmonious relationship with our planet.

I sincerely hope that this Mushroom Farm progress guide has given you more assurance in your capacity to grow mushrooms. Throughout your adventure, you can refer to this guide and my actions if you have any questions. However, be aware that there are numerous effective approaches. So start growing—you've got this!